SQUIRRELS

BRIAN WILDSMITH

It is easy to recognize a squirrel. He is
a furry, small animal with a long, bushy
tail, two strong back legs, two small

front paws, two large tufted ears which
stick up, and two big front teeth. He
looks happy and mischievous.

In summer-time the squirrel's coat is quite thin. But in winter-time it grows thick and

strong. He seems to have little socks on his
feet and warm fur-gloves on his front paws.

Squirrels live in trees.
Sometimes their home is
a hole in an old tree-trunk.
Sometimes it is high up
in the tree top in an
abandoned crow's nest.

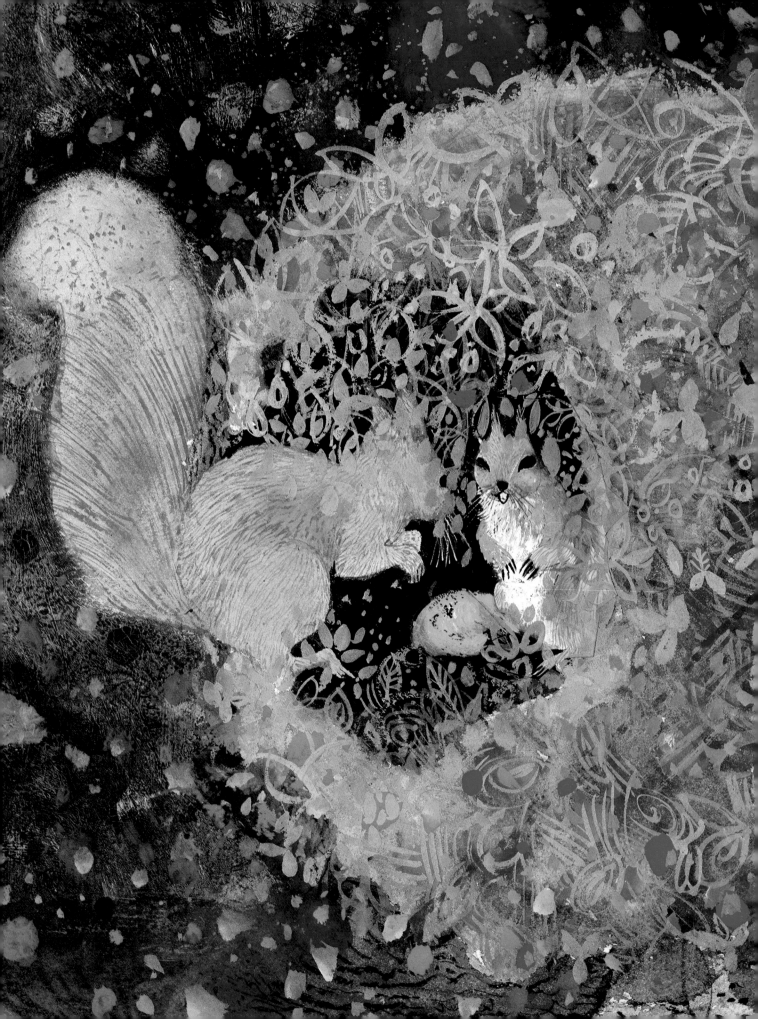

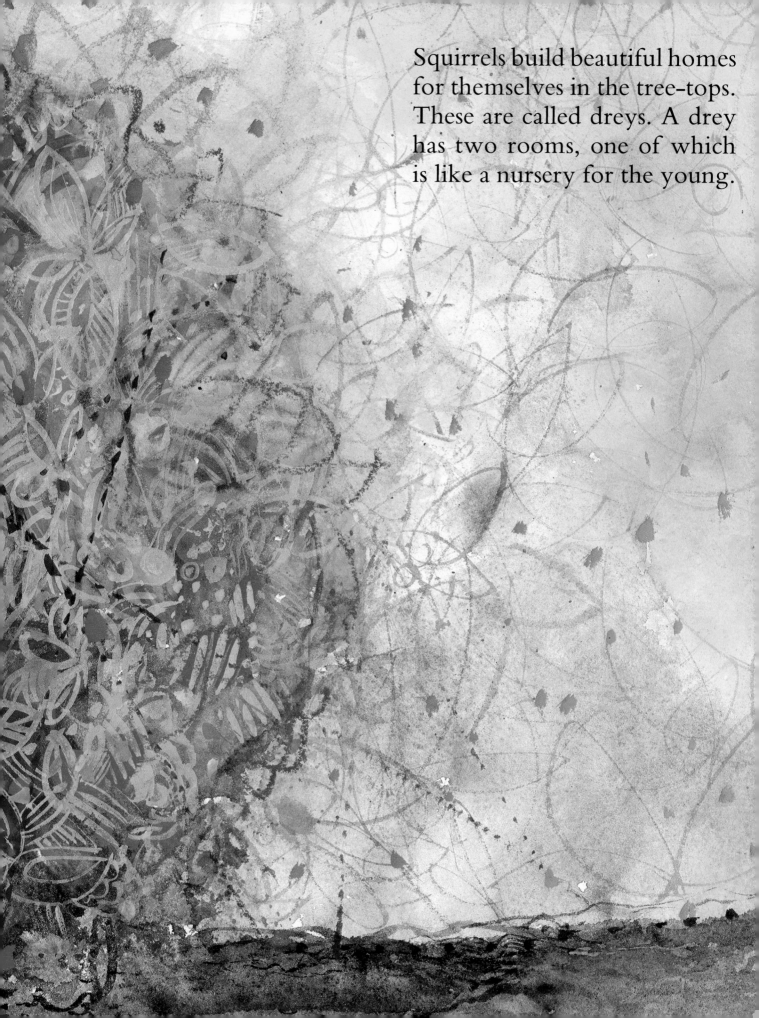

Squirrels build beautiful homes for themselves in the tree-tops. These are called dreys. A drey has two rooms, one of which is like a nursery for the young.

Their sharp claws give them a strong
hold on the trunk of a tree, so that they

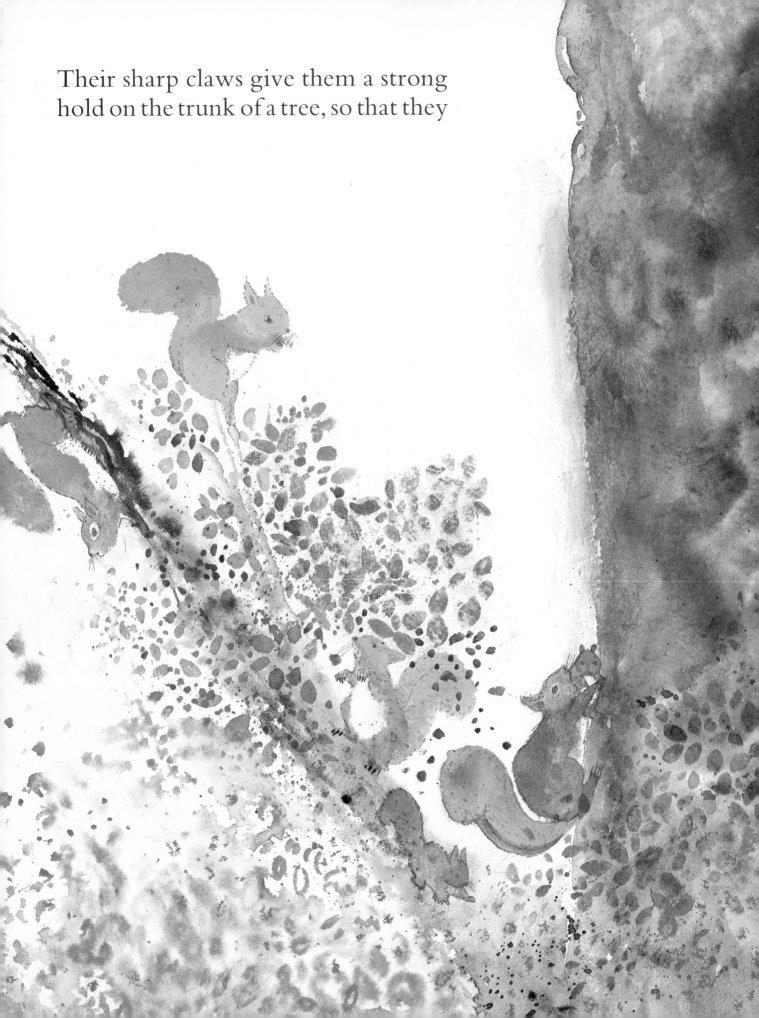

can scamper both up and
down it with great speed.

Probably no animal in the world uses his
tail for so many different purposes. When a
squirrel leaps through the air from tree to tree,
he can use his tail as a parachute, and it even
helps him to change direction.

And when a squirrel swims, as he does sometimes, he can use his tail as a sail.

When he scurries along the bough of a tree,

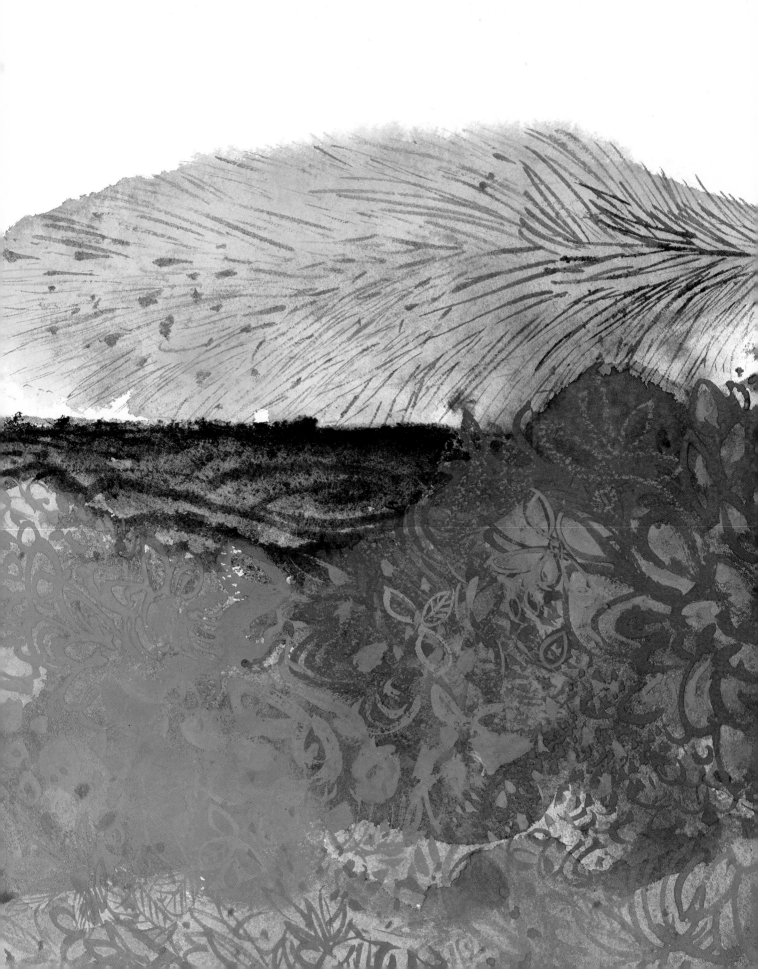

he can use it to balance and guide himself.

And when a
squirrel sleeps,
he wraps his tail
round himself
like a blanket.

Twice a year – in the spring and at the end of
the summer – the female squirrel has her young,
usually three or four babies at a time. For seven

weeks the mother gives them milk, then they
are ready for a squirrel's favourite food – nuts,
acorns, wheat, fruit.

Sometimes squirrels will steal birds'
eggs, and they are very fond of mushrooms.

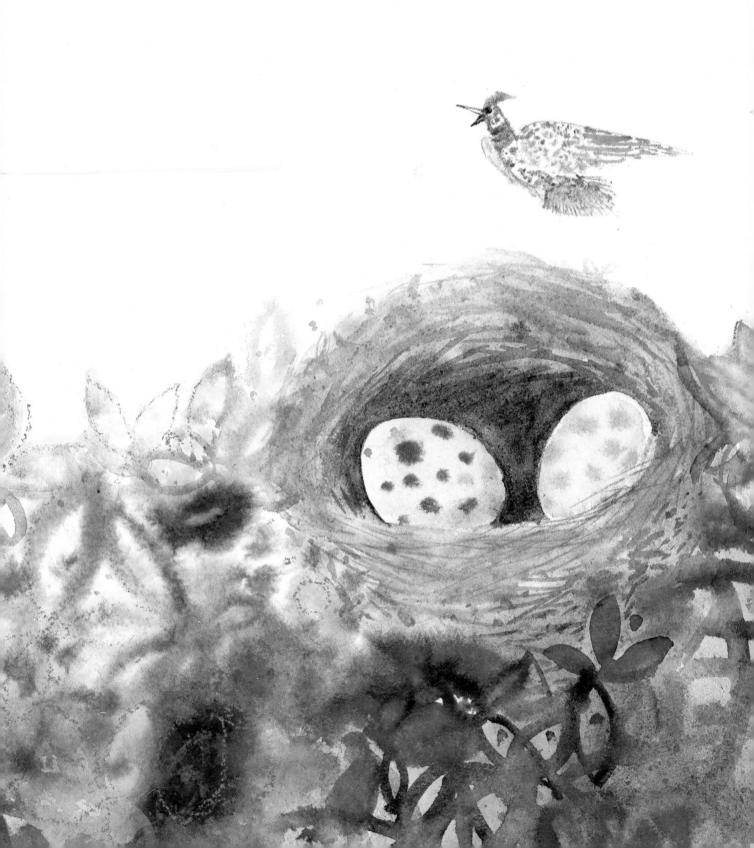

In the last
weeks of autumn,
squirrels become
very busy
gathering nuts
and acorns
to eat during the
winter. They
carry them in
their cheek
pouches and hide
them in all kinds
of nooks and
holes roundabout
the trees where
they live.

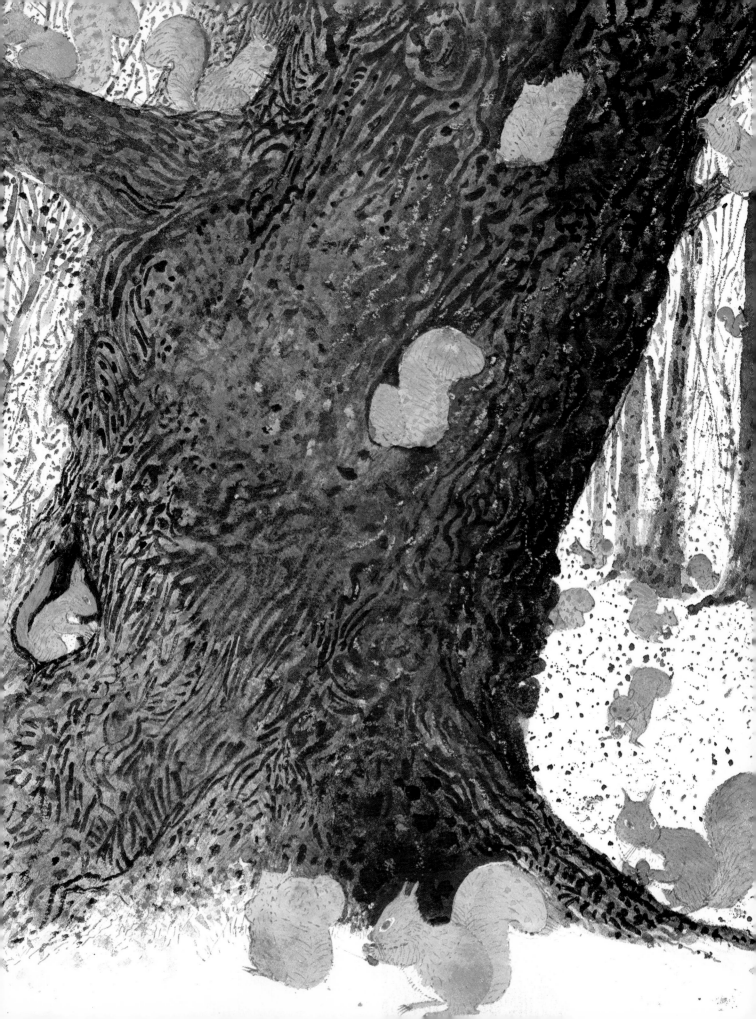

Later on they come back to look
for them, but quite often they

forget where they have hidden their stores and
dig in several places before they find them.

Squirrels are delightful little creatures, but
they can be destructive. In a garden they

like to pull up new plants, eat bulbs and
even bite the tops off very young trees.

Next time you walk in the woods, if you are quiet
and observant, you will perhaps see squirrels contented
and busy. They might be jumping from tree to tree,
frolicking on trunks and boughs, or possibly hiding a
store of nuts for the winter.